AF228920

# Satellites

by Julie Murray

Dash!

LEVELED READERS

3

An Imprint of Abdo Zoom • abdobooks.com

**Level 1 – Beginning**
Short and simple sentences with familiar words or patterns for children who are beginning to understand how letters and sounds go together.

**Level 2 – Emerging**
Longer words and sentences with more complex language patterns for readers who are practicing common words and letter sounds.

**Level 3 – Transitional**
More developed language and vocabulary for readers who are becoming more independent.

## abdobooks.com

Published by Abdo Zoom, a division of ABDO, PO Box 398166, Minneapolis, Minnesota 55439. Copyright © 2020 by Abdo Consulting Group, Inc. International copyrights reserved in all countries. No part of this book may be reproduced in any form without written permission from the publisher. Dash!™ is a trademark and logo of Abdo Zoom.

Printed in the United States of America, North Mankato, Minnesota.
102019
012020

Photo Credits: Alamy, Getty Images, iStock, NASA, Science Source, Shutterstock
Production Contributors: Kenny Abdo, Jennie Forsberg, Grace Hansen, John Hansen
Design Contributors: Dorothy Toth, Neil Klinepier, Victoria Bates

### Library of Congress Control Number: 2019941335

### Publisher's Cataloging in Publication Data

Names: Murray, Julie, author.
Title: Satellites / by Julie Murray
Description: Minneapolis, Minnesota : Abdo Zoom, 2020 | Series: Space technology | Includes online resources and index.
Identifiers: ISBN 9781532129285 (lib. bdg.) | ISBN 9781098220266 (ebook) | ISBN 9781098220754 (Read-to-Me ebook)
Subjects: LCSH: Satellites--Juvenile literature. | Space surveillance--Juvenile literature. | Space sciences--Juvenile literature. | Technology--Juvenile literature. | Astronautics--Juvenile literature.
Classification: DDC 629.4--dc23

# Table of Contents

# Satellites

Look up in the night sky. Do you see a bright light moving in the distance? It may be a satellite. There are about 5,000 satellites currently in **orbit**!

A satellite is an object that **orbits** around another object in space. It can be a **natural** satellite like the Moon. It orbits around Earth. Others are **artificial** satellites. They are man-made and sent into space.

First
Satellites
8

The first **artificial** satellite was Russia's Sputnik 1. It launched in 1957. It was the size of a beach ball. It sent signals back to Earth for 22 days.

EXPLORER 1
AMERICA'S FIRST EARTH SATELLITE
jpl
JET PROPULSION LABORATORY
CALIFORNIA INSTITUTE OF TECHNOLOGY
10

The first US satellite was Explorer 1. It went into space in 1958. Explorer 6 was the first satellite to send back a picture of Earth in 1959.

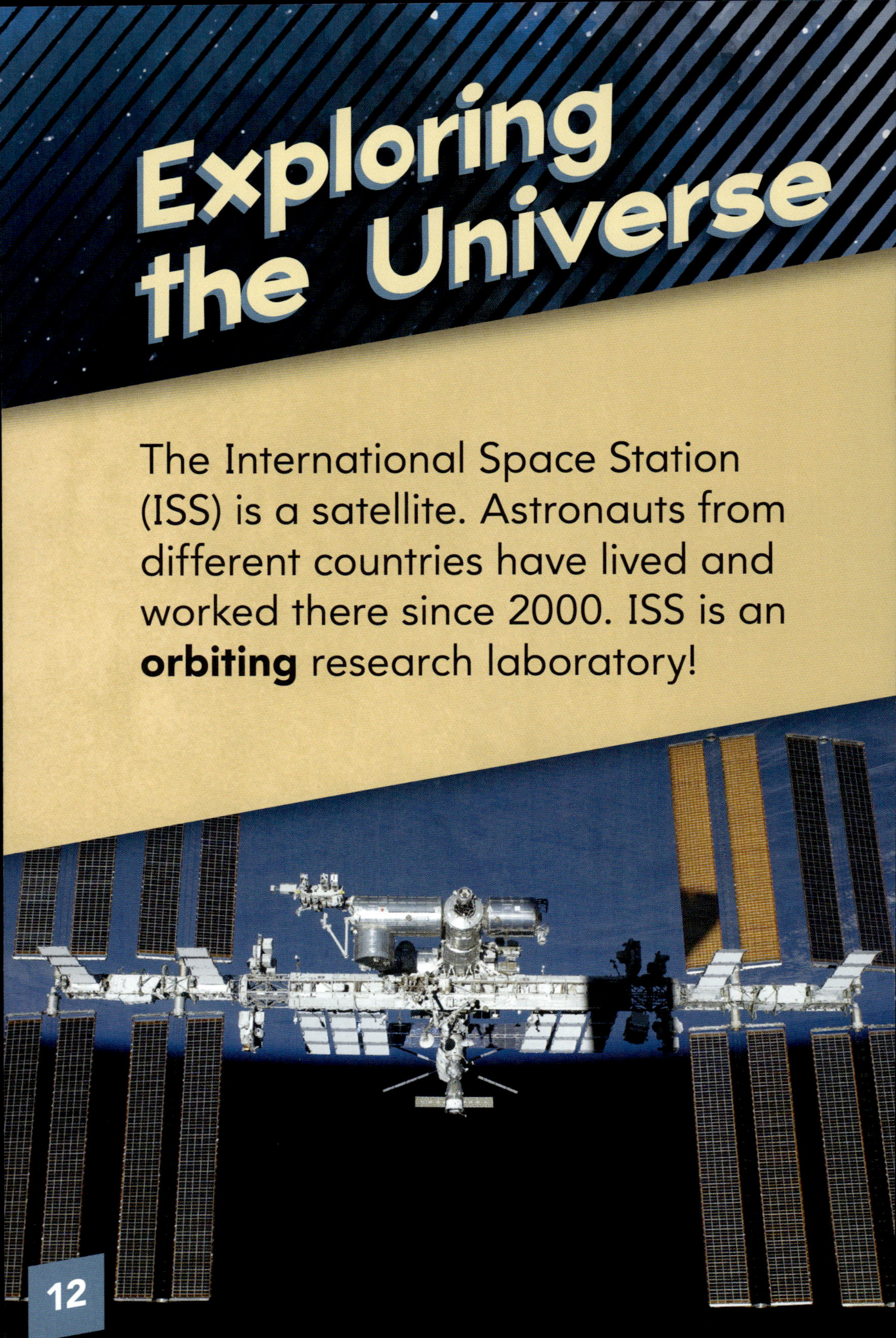

# Exploring the Universe

The International Space Station (ISS) is a satellite. Astronauts from different countries have lived and worked there since 2000. ISS is an **orbiting** research laboratory!

The Hubble Space Telescope has been in **orbit** since 1990. It has traveled more than four billion miles! Hubble has captured more than 1.3 billion photos. These photos have changed our understanding of the universe.

Cassini was a satellite that explored Saturn. It **orbited** the planet for 13 years. It sent back many images of Saturn's rings and moons. It carried the *Huygens* probe. This probe landed on Titan, Saturn's largest moon.

We rely on satellites every day. Global Positioning System (GPS) is made possible by satellites. We use GPS for driving directions and finding the closest store. It is also used to **navigate** planes and boats.

Satellites are used to track the weather and watch TV. We also use them for the internet and to talk on the phone. They are an important part of our lives each day.

TV PAUSE
TITLE LIST
REC
SOCIAL VIEW
FOOTBALL
DISCOVER
APPS
SEN
RETURN
BACK
DIGITAL/
ANALOG
EXIT
1
4
7
TV
+
−

# More Facts

- Yvonne Brill was a rocket scientist. She invented a rocket thruster. This keeps satellites from slipping out of their **orbit**.

- The Hubble Space Telescope has led to the discovery that the universe is about 13.7 billion years old. It has also revealed thousands of other galaxies.

- Some satellites are only a few hundred miles off the ground. Others are more than 22,000 miles (35,406 km) away!

# Glossary

**artificial** – made by human beings; not natural.

*Huygens* – a probe, named after the Dutch astronomer who discovered Saturn's largest moon Titan in 1655, that landed on Titan in 2005 to send photographs and data about the atmosphere to scientists on Earth.

**natural** – made by nature, not humans.

**navigate** – to travel along or through.

**orbit** – (n) the curved path in which a natural or artificial body moves in a circle around a star, planet, or moon. (v) to move in a circle around.

# Index



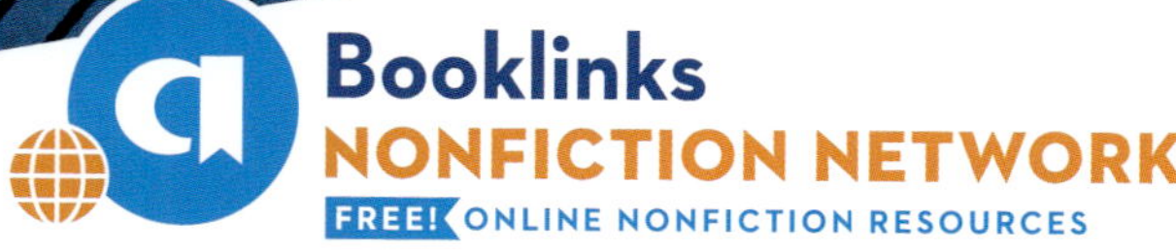

To learn more about satellites, please visit **abdobooklinks.com** or scan this QR code. These links are routinely monitored and updated to provide the most current information available.